Evidence for Evolution

Evidence for Evolution

Noel Babb

VANTAGE PRESS
New York

FIRST EDITION

Published by Vantage Press, Inc.
516 West 34th Street, New York, New York 10001

Manufactured in the United States of America
ISBN: 0-533-10775-X

0 9 8 7 6 5 4 3 2 1

To Frances, my wife of fifty-six years,
whose love, understanding, and encouragement
made this book possible

Contents

Foreword

A recent Gallup poll taken in 1991 shows that 47 percent of the people of the United States believe that mankind was created as he is today in a single act of creation and within the last ten thousand years.

Many of our schools from college level down do not teach evolution in their science rooms; in some biology textbooks, the word *evolution* is not even mentioned; and many science writers are too technical and complex for the average reader. It seems to me there is a need for someone to write a simple summary of evolution.

I have attempted to put in simple language some of its basic characteristics so that one might better understand.

Noel Babb
Poteau, Oklahoma
November, 1991

Acknowledgments

To Eugenia C. Scott, publisher of *Creation/Evolution* and executive director of Nation Center for Education, who helped me with science content of this book.

To Elizabeth Watson for her help with punctuation, correct English, etc.

To Carol Baker, printer, who made transcripts and without whose help this book would have been impossible.

To Laneal, my daughter, and her husband, Larry, who encouraged me to publish.

To my wife, Frances, and all my friends and relatives who helped.

Noel Babb
June 1993

Evidence for Evolution

Some Definitions

ORGANIC EVOLUTION. Descent with modification. The idea that living forms shared common ancestors in the past, from which they have changed. This means that an offspring is not always exactly like its parents. Sometimes there may be a mutation that helps an organism better suit its local environment. There can also be a change in the arrangement of its genes. The new form's chances of survival might be better than those of its parents.

SPECIES. A group of related organisms that have certain characteristics in common and that interbreed freely.

GENE. The part of a cell that is passed to offspring and which causes it to have certain characteristics, sometimes in a different form (a mutation). Each gene influences the inheritance and development of some characteristic of an organism.

FOSSIL. A part or trace (the track of an animal or impression of a leaf, etc.), generally buried in a layer of earth. It can last for vast periods of time.

STRATA. Layers of mud, sand, or dirt washed down from higher ground, laid down horizontally ages ago and compressed by the weight of the earth on top. This is where most fossils are found. In the Grand Canyon, these strata, or layers, are a mile deep. Some strata on the earth are not horizontal anymore; the crust of the earth has moved, tilting them.

GEOLOGY. The science that deals with the earth, the layers of which it is composed and their history.

PLATE TECTONICS. The theory that the earth is divided into plates that move.

NICHE. The place and function of an organism within an ecosystem. CREATIONISM. The idea that the earth is a few thousand years old and that all life-forms were created as you see them today.

Explanation of Evolution

Two hundred years ago, I suppose it was natural for people to think that life-forms were always pretty much as they saw them. After all, no one saw one life-form change into another, and they knew very little about fossils. We know today that species have changed. Go to any good museum of natural history and you can see that the skeletons of animals of many millions of years ago are very different from the skeletons of animals of today.

We do have descent with modification: life-forms or species have changed, as Darwin told us. Mainline scientists consider evolution a fact not a theory anymore. Look at our national parks and monuments. At Petrified National Park in Arizona are the skeletons of animals and the petrified remains of trees that are different from those of today, similar and in some cases ancestral, but different enough according to paleontologists to be classified as different species or life-forms. This is true also of fossils at Agate Fossil Beds in Nebraska, Dinosaur National Monument in Utah, John Day Fossil Beds in Oregon, and many others.

The evidence is there, the fossil record is clear! Species have changed!

There seem to be two patterns of evolution: (1) A gradual change in one life-form from one thing into another. An example might be mankind. From *Homo habilis* to *Homo erectus* to *Homo sapiens*, a species can be altered so extensively by natural selection as to be changed into a different species. Yet no matter how much change occurs, only one species remains. (2) Due to canyons,

3

mountains, etc., or by intrinsic isolating mechanisms, different groups of animals are kept apart and do not interbreed so that they split and through time establish new species. Each breed goes its own way and does not mix with others.

More about the Nature of Evolution

Punctuated Equilibria

Charles Darwin believed evolution took place gradually; most life-forms changed very, very little at a time. The rate was mostly very slow and the reason we did not find any more transitional forms was the imperfection of the fossil records. This view has changed.

Niles Eldredge and Stephen Jay Gould came up with a theory that helps explain this apparent mystery. According to their theory of punctuated equilibria, the rate of change in populations is not steady but episodic. There can be periods of quick change, geologically speaking, and long periods of stasis, maybe millions of years when forms stay the same, short in a geologic sense. This is one way of accounting for the "gaps" in the fossil records. Darwin thought these gaps were due to our not knowing enough about the fossil record. Now we think conditions could change in hundreds or thousands of years, due to climate, glaciers, etc., but then remain the same for millions of years. If conditions were mostly the same, there would be no reason for forms to change. These conditions could last millions of years.

What should the fossil record show? Just what you find! There are few transitional forms because most of the life of populations is spent in a state of stasis where no change is necessary.

Progress in Evolution

Did life begin with very small organisms and progress with man as the ultimate goal? It seems to be hard for Western man to think otherwise. Darwin told us to be careful about calling one species superior to another. Does a man's brain better prepare him for his environment than a fish's fin, a bird's wing, or a deer's swiftness? I think not. They seem to fit their surroundings as well as we do. We all seem to be subject to natural selection.

The fossils from the Burgess shale of some 530 million years ago seem to show us that life did not start with us in mind but developed many basic life-forms from which a lesser number were later selected by nature and probably by chance. These became the many diversified forms of today. Play the tape of life over again and man might not be here.

So maybe all things were not put here for our pleasure and benefit, and man is just a part of nature. Does believing this make us evil or less moral? A better understanding could encourage us to create a better world.

Charles Darwin 1809–1882

Upon completing his B.A. degree in 1831, Darwin, at age twenty-two, signed on as naturalist aboard the British ship, the *Beagle*. They sailed along the coast of South America, visited many islands and other lands. The voyage lasted five years.

When he left England, Darwin, like most of the people of the world, was a firm believer in the fixity of species or special creation. But after visiting many lands, especially the Galapagos Islands, where he saw animals similar to the ones he had seen in South America 450 miles away, yet different enough to be different species, he could not keep from wondering about what had caused them to change.

Darwin was a keen observer, very intelligent, and a hard worker. Other scientists helped him also in determining that especially the three mockingbird kinds found there on the Galapagos were three different species, even though they looked somewhat like the mockers from South America. After much thought and study, he finally came up with the theories of evolution and natural selection.

In 1859, he wrote his great book, *Origin of Species*. By 1869, most scientists had accepted the theory of evolution it proclaimed.

Darwin refuted the belief in the individual creation of each species, establishing in its place the concept that all life descended from a common ancestor.

Geologic Time Table
(Not to Scale)

We are part of a vast continuum of life stretching back billions of years.

Era	Period	Epoch	Major Events
Cenozoic Last 65 million years	Quaternary	Recent	Development of human agriculture.
		Pleistocene	Successive ice ages. *Homo sapiens* widespread
	Tertiary	Pliocene	Human evolution.
		Miocene	Families of mammals arise; others die out.
		Oligocene	Horses and rhinoceroses, dogs and cats came into existence.
		Eocene	Most modern orders of mammals distinguishable.
		Paleocene	Great diversification of primitive mammals.
Mesozoic Age of Dinosaurs 180 million years	Cretaceous		Major extinctions, including last dinosaurs. Continents fairly well separated.
	Jurassic		Dinosaurs diverse. First birds and first traces of flowering plants.
	Triassic		Continents begin to separate. Pinelike plants (gymnosperms) dominate. Some primitive mammals.
Paleozoic Ancient Life 325 million years	Permian		Diverse reptiles aggregated into one.
	Carboniferous		First reptiles.
	Devonian		Age of fishes. First amphibians and insects.
	Silurian		First plants and arthropods.
	Ordovician		First jawless fishes.
	Cambrian (570 million years)		Appearance and rapid diversification of most animal phyla.
Precambrian 3,990 million years			Origin and diversification of algae.

Explanation of Timetable

Notice that the Precambrian Age is over eight-ninths of the earth's age.

These great periods of time mean something; they are not just names. At the end of each period, a great many species died off. This was due to changes in the environment.

Something caused conditions to change. Many forms of life could not change and became extinct. Other forms took their place.

Evolution takes long periods of time to work. Most people thought the earth was only a few thousand years old until Darwin's book *Origin of Species* was published. Darwin knew there had to be very long periods of time in order for his theory to be true. This has proved to be the case.

Evidence for Evolution

Why Should We Study Evolution?

1. Grand Canyon: Up to one mile deep and eighteen miles wide. How can we believe this only took a few thousand years to form? The fossils found in it are from life-forms millions of years old. Just think of the vast amount of time necessary for these fossils to form and be buried this deep! Why aren't mammal bones found fossilized in it? Because, of course, mammals arrived on the scene many millions of years later.

2. Dinosaur National Monument: How can we understand about the dinosaurs without knowing something about when they lived, how they lived, and when they died? Evolution helps explain this.

3. Petrified Forest: You have to allow for vast periods of time for these trees to have lived, died, and become petrified (turned to stone). Scientists tell us also that these trees were of different species from those of today.

4. Agate Fossil Beds in Nebraska: How are we going to account for bones of extinct animals, 20 million years old, unless we know about evolution?

The above are just a few examples: There are many more.

Maropus: A horselike animal from the Agate Fossil Beds in western Nebraska. The bones are large. Notice the claws on its feet! (Fig. 1).

Fig. 1. Maropus. From the Miocene epoch some twenty million years ago.

Why Life Changes

1. The English Moth

In English forests there is one species of moth with both white and dark individuals, living on trees. Before the Industrial Revolution in England, most of the English moths were white because the bark of the trees they rested upon was white; the birds could not see them very well and ate more dark ones. Birds acted as natural selectors, so there were more white moths over time. Then soot from the factories in the Industrial Revolution caused the bark of the trees to turn dark, so moths that were dark were more apt to survive. The birds could not see them as well. Most of the moths became dark.

These kinds of quick changes that happen in a few years can be quite frequent. Insects that survive certain poisons can cause new varieties of the same species to develop. Members of a population resistant to the poison have more offspring than those that are not resistant, and therefore the genes for resistance spread through the population. That makes crop protection difficult for farmers.

2. The Domestication of Animals

When Darwin wrote his book *Origin of Species*, he began by showing how man has acted as the selector in devloping animals from the wild into tame ones that he desired for his own use.

There are many examples:

(a) The Dog: Just think! From the wild dog came hunting dogs and guard dogs. In modern times have come all the other kinds. These came about through selection; people kept and bred those that came nearer to suiting their purposes.

(b) The Cow: Look at the wild cow that gave enough milk for

one calf, and then the dairy cow that gives enough for several calves. People brought this about through selection; they knew what they wanted and kept and bred those best suited for their purpose. Living things do not always stay the same.

These are examples of people acting as the selector. Nature can also act as selector. Life best suited to its local environment has the best chance of living and leaving offspring. Those that can't make the change become extinct. Their absences leave new niches that are eventually filled by new life-forms.

How Life Changes

Another example of branching or population splitting: One, this time due to the Grand Canyon. Two populations of squirrels separate, each going its own way. Gene pools of both populations are not mixed.

The Kaibab Squirrel

The Grand Canyon has been a barrier to some animals. One notable example is a tassel-eared squirrel—the Abert (not Albert) squirrel, found on the south rim. It is light brown with white underparts. A close relative, the Kaibab squirrel, is rare and is found only on the north rim of the canyon. It sports a bushy white tail and dark gray body.

Both mammals are dependent on the Ponderosa Pine for food, and they cannot live where it doesn't grow. Biologists think that when the canyon was cut the squirrels developed into subspecies because they could not cross the desert or the Grand Canyon, which was created by the Colorado River. The two animals have been kept separate for several thousand years and have already started to change.

Abert squirrel   Fig. 2.   Kaibab squirrel

This is an example of evolution taking place; when animals can't inter-breed, natural selection, genetic drift, or both causes them to develop in their own ways. It is also an example of specialization: two species forming from one.

An Example of Punctuated Equilibria

The existence of pupfish in Death Valley is an evolutionary miracle. Survivors of Ice Age lakes, they found refuges where they could evolve into several species that live in fresh and salt waters, ponds and streams. The cottontail marsh pupfish live in the saltpan, a habitat saltier than the sea. These fish were isolated for thousands of years and evolved into new species that could exist for millions of years.

Transitional or Intermediate Forms

Part Reptile, Part Bird: Archaeropteryx

Archaeropteryx was part bird and part reptile. It lived in the Jurassic Period, some 140 million years ago.

14

Fig. 3. Archaeropteryx

A fossil of Archaeropteryx was found in a block of limestone in 1881. It shows a stage of development from a reptile to a bird. Four more specimens have been found.

There are impressions of feathers in the rock surrounding the specimen. It had clawed fingers on its wings and a long, bony tail like a reptile.

Part Ant, Part Wasp

Resin sap from certain trees is sticky and catches insects sometimes as they rest on bark. It can later change into amber, which can then last millions of years.

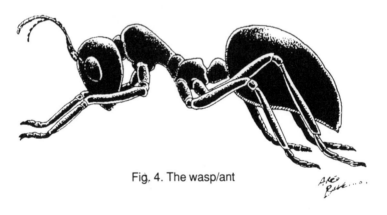

Fig. 4. The wasp/ant

The insect in Figure 4, found in a piece of amber, is part wasp and part ant. Amber is fossilized vegetable resin, like sap from a tree. The wasp/ant lived during the Cretaceous Period, 100 million years ago.

The Story of the Horse

This story shows the horse changing from eophippus to equus (horse of today).

Old biology textbooks used to show the evolution of the horse from little eohippus right up to the modern horse. This is not the case. Figure 5 shows that there were many side branches to the tree; most forms of horse adapted to the local environment and later became extinct.

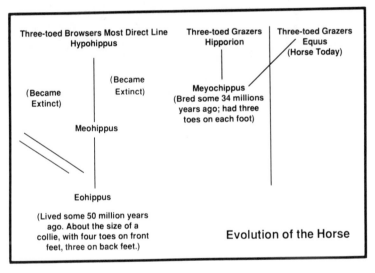

Fig. 5

Figure 6 shows how the horse's foot and leg have changed through time. The one-toed horse could run faster across the plains and get away from his enemies more easily.

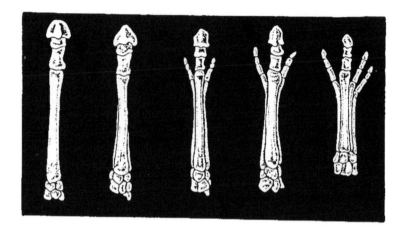

Fig. 6. Development of the horse leg and foot

The evolution of the leg and foot of the horse changed over time. This did not occur at a steady pace. Some species of horses that later became extinct had four, three, or two toes. The modern horse is not descended from the main branch but is a twig on a side branch of the horse line.

Other Transitional Forms

Dimetrodon: Part Reptile, Part Mammal—from the Permian in Texas

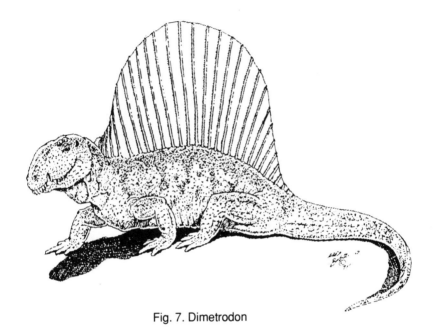

Fig. 7. Dimetrodon

The dimetrodons were powerful meat-eaters, up to three and a half feet long. They had a deep, narrow skull, with very large teeth used for attacking other animals and tearing their flesh. The sail on its back probably helped the animal regulate its body heat.

The change from reptile to mammal is evident mostly in its teeth. The teeth showed a change from reptile to mammal.

Below is a picture of some skulls displayed in the Little Museum of Natural History, Poteau, Oklahoma.

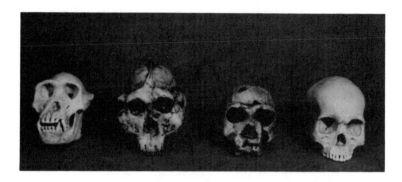

Fig. 8

On the left is the picture of the skull of a chimpanzee. To the right of it is *Austropithecus boisei*, then *Homo erectus*, and last, to the far right, is *Homo sapiens*, modern man. I think anyone can see the similarity of the four skulls, but there are obviously many differences too. The entry of the spinal column to the head is directly below on animals that walk erect, but sits at an angle on four-footed animals. It is thought the chimp split off the primate branch some five to seven million years ago; *Austropithecus boisei, Homo erectus,* and *Homo sapiens* much later.

Notice the crest on the skull to which strong jaw muscles were attached. These muscles and the large teeth would help chew through leaves and other vegetation. *A. boisei* became extinct.

He probably ate grubs, roots, and meat from other animal kills.

Notice that transitional forms do not have to be ancestral but can branch off, go their own way, or even become extinct. Some fossil species similar to Australopithecine are those of our ancestor,

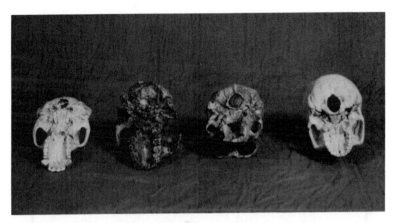

Fig. 9

even if most of the known fossils were not. Sort of like me and my cousin. I live at the same time as my cousin, but I am not descended from him. We share a grandfather, perhaps like the gracile and

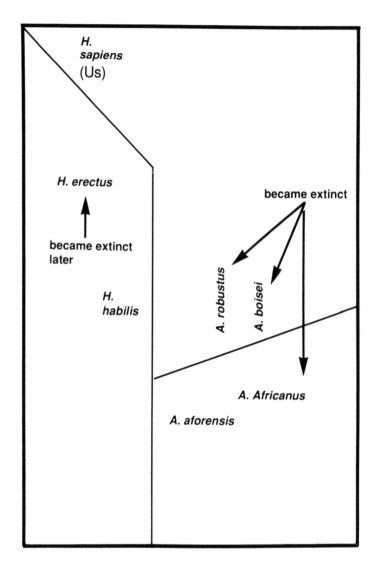

Fig. 10

robust Australopithecines shared an ancestor and were not descended from one another. On the other hand, I look more like my cousin than I look like you because he and I shared a common ancestor in the recent past. When we were children, our grandfather was still alive, so we would have been coexisting with our common ancestor for a while (as *H. erectus* lived at the same time as the latest Australopithecines). The fossil record can show species that gave rise to later species, coexisting with them for a while and then dying out.

Figure 10 shows *Austropithecus boisei*, *A. robustus*, and *A. africanus* all becoming extinct. They each went their own way and later died out while *Homo erectus* became our ancestor.

Chemical Basis of Heredity

Recently in the field of biology, there have been important discoveries as to the exact manner in which offspring receive traits from their parents. This new knowledge provides abundant evidence of ancestral genealogies—a basic test of evoltion. The cell nucleus contains the genetic machinery of life, the genes and the chromosomes in which the genes are packaged. Genes are the ancestral units of heredity—blueprints for growth development and instinctive behavior. Maps of the gene sequence provide a mathematical key to ancestral history—that is, evolution.

Molecular biologists have demonstrated that the genetic molecules, although complex, possess a simple code that is the same for all organisms. This is strong evidence that the present life on Earth originated only once, in a single kind of organism. The magnificent diversity that we now see about us stems from that single ancestry.

Darwin knew life changed, but he knew nothing about genes. Genes are units of heredity and are passed from generation to generation. Sometimes they are modified by a change in their structure or a change in their arrangement. This is called a mutation. If the surroundings change enough, life has to change or perish.

Convergence

To converge means to come together, to meet at a point. Convergent evolution is the appearance of similar characteristics in organisms not closely related to one another.

We might think, well, what does this have to do with evolution or natural selection? I have chosen the examples of the Australian wolf of Australia and the American wolf of North America. They are two entirely different lines or types of animals; the Australian wolf is a marsupial, the American animal a placental—an entirely different line thousands of miles away. But they are both carnivores, both occupy similar niches in their environments, and both have developed, due to the environment, similar features. Why is this?

I think this is a good example of natural selection at work. There was pressure for an animal to fit its environment, and since the environments were similar, the features of the animals became similar. Look at the drawings and this becomes clear.

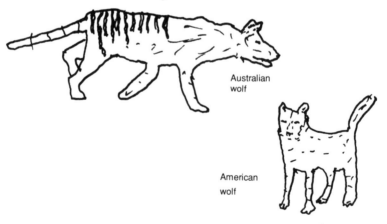

Australian
wolf

American
wolf

There are many other examples of convergence.

Embryology

During the fourth week, the human embryo becomes recognizable as a mammal. It becomes C-shaped and a tail is visible. The umbilical cord forms and the forebrain enlarges.

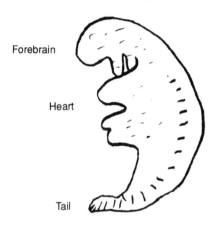

Forebrain

Heart

Tail

The embryo goes through several stages: when it has (1) gill slits, (2) extensive hair on it, (3) a tail.

Now, why do you suppose that all of us have these animal characteristics before birth? It only makes sense to me if we are kin to animals, if we all had common ancestors.

This is more evidence for evolution.*

*This data comes from three books I have read: Haldane, *The Causes of Evolution*; Ruse, *Darwinism Defended*; and *The Encyclopedia of Medicine*, 1989.

Vestiges

A vestige is a part or organ of the body that is no longer fully developed or useful but performs a definite function in an earlier stage of the existence of an organism. Rudimentary organs include the vestige of a horn in hornless breeds of cattle or vestige of an eye in cave animals. Vestiges in the human body include the appendix, the coccyx, the wisdom teeth, nipples on males, and muscles to move the ears in some people. Also, some snakes have vestiges of legs that, of course, are no longer useful.

How can we account for these useless parts? No, they make sense only if we consider them no longer useful in their present local environment. Life-forms adapt, as evolution proclaims.

Isolation

Through isolation new species arise. There are two causes of isolation:

1. Geographic barriers, where features of the earth's surface, such as mountains, canyons, etc., separate populations and allow them each to go their own way without interbreeding. Each gene pool is kept separate until a new species is formed.

Darwin, as you know, found new species: birds, reptiles, etc., on the Galapagos Islands. These life-forms were different from those of South America, 450 miles away. His final conclusion from this problem was that birds, reptiles, etc., migrated from South America to these islands in the distant past and changed to fit the new environment—becoming new species.

This same thing happened probably also in New Zealand, the Hawaiian Islands, Australia, and other places that were isolated from mainlands. You see, these are examples of new species being formed when small populations are isolated due to some barrier and kept free from other groups of the same organisms. This same situation could result if the barrier was a mountain range, canyon, etc., just so the populations are kept apart and not allowed to interbreed.

2. Intrinsic isolating mechanisms. By intrinsic is meant hereditary. In other words, differences prescribed by the genes of the opposing populations come into being and each population goes its own way, kept separate by not interbreeding. Somthing happens step by step through the reproductive process to keep them apart until a new species is formed. For instance, certain flycatchers, little birds that dart out from tree limbs to snatch flying insects, consist of five species and occur together in the northern United States. They remain genetically distinct in part because they prefer different habitats and partly because each uses its own identifying call during breeding season.

Another example of intrinsic isolating mechanisms is found in the giant silkworm moth of North America, which fly and mate at various times during late afternoon. The females call in the males by secreting a powerful chemical scent. This attracts only males of her species, and although there are sixty-nine different species in the same region, each one is kept distant by this reproductive process.

Homology

The drawing below shows that the bones of a crocodile's front leg, a human's arm, and a bird's wing are all similar. Now why, do you suppose? The only way this makes sense to me is to recognize that somewhere back down the line they had a common ancestor.

To an evolutionist, homologies are expected and they form a bright thread in the evolutionary fabric. Adaptively valueless similarities point to the fact that widely different organisms are descended from common ancestors. Natural selection has taken the ancestral form and molded it to different ends.

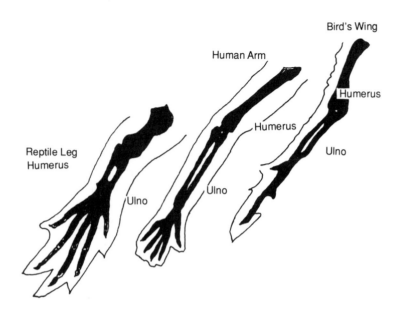

Bird's Wing

Human Arm

Humerus

Humerus

Ulno

Reptile Leg
Humerus

Humerus

Ulno

Ulno

Ulno

The panda's thumb is an example of a basic design that had gone too far to change, so its wrist bone was modified to fit the new circumstance; the five digits of its paw could not change, so its wrist bone was changed instead to make the sixth digit. This was an example of a carnivore bearlike animal becoming an eater of plants. The new digit developed from the wrist bone, not perfectly, but the animal got by. Nature can take a basic form and modify it to fit a different environment, when it can't make the basic form again.

Extinction

Earth has had five major extinctions and several minor ones, the latest major one coming at the end of the Cretaceous, when the demise of the dinosaurs occurred.

Something in the environment changes. Certain life-forms cannot change, so they perish, creating new vacancies to be filled by new species. The fossil record shows this has happened many times. Skeletons of animals and plants that lived many millions of years ago are unlike others that lived later. Now, were new creations made each time this happened or did new forms evolve to fill these vacancies? Scientists say evolution occurred.

According to plate tectonics, the crust of the earth is in constant motion. Although very slow, it moves. Also volcanic eruptions, comets, impacts, etc., can create new environments and if a species cannot change or move to a favorable environment, it becomes extinct. They tell us 99 percent of all life-forms that ever lived have perished. *Homo sapiens*, mankind, will someday follow suit.

Kinship

Now, we come to the question of all questions, how did we, mankind, get here? Were we created all at once in our present form, some ten thousand years ago? According to a Gallup Poll of 1990, some 47 percent of us think so. Let us look at the evidence.

First we have now hundreds of manlike fossils that show we have evolved from one form of primate to another form. There are not just a few, but many, many bones to show that this happened and is accepted by some scientists as true. The earliest bones are from an apelike form, Lucy, who lived some three and a half million years ago, and is called *Australopithecus afarensis*. She had features of both ape and man. More involved specimens are found, from *Homo habilis*, more manlike, to *Homo erectus*, to us, *Homo sapiens*. More of these bones are being found all the time. The human tree now looks something like what we see in Figure 9.

Bones are not the only thing that show our relationship to other primates. So do genes. Scientists tell us our genetic makeup is 96 percent like the chimp. We also have muscle for muscle and bone for bone like the chimp. We both use tools, as discovered by Jane Goodall. She discovered chimps take a limb or branch of a tree, trim it to suit them, and secure termites from a hill. She also discovered many other humanlike qualities in them. This does not mean, of course, that man and chimp are equal, but it does show kinship.

Also, what about the liver transplant shown on television in which doctors took the liver of a baboon, transplanted it in a man, and he lived for several weeks! Why did they not use the liver of a calf or reptile? Because the baboon was a primate and more closely related to man, I think.

Man is a part of nature, not apart from it. Why can't we accept it?

We do not accept the American way. If we were on a jury when they were trying a man for his life, would we condemn him before the evidence was in, even though we might not like him? Surely not!

I think we would hear the evidence, then make up our minds. Why can't we do this about evolution! There are mountains of evidence for it! Why don't we accept it?

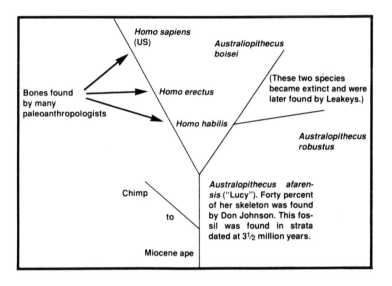

Homo sapiens
(US)

Australiopithecus
boisei

Bones found
by many
paleoanthropologists

Homo erectus

(These two species
became extinct and were
later found by Leakeys.)

Homo habilis

Australopithecus
robustus

Chimp

to

Miocene ape

Australopithecus afaren-
sis ("Lucy"). Forty percent
of her skeleton was found
by Don Johnson. This fos-
sil was found in strata
dated at 3½ million years.

References

Books

British Museum Department of Paleontology Staff. *The Prehistoric Age*.

Darwin, Charles. *Origin of Species*.

Dobshansky, Theodosius. *Genetics and the Origin of Species*.

Eldredge, Niles. *Fossils: The Evolution and Extinction of Species*.

———. *Time Frames*.

Futuyma, Douglas J. *Science on Trial*.

Gould, Stephen J. *Ever since Darwin*.

———. *Wonderful Life*.

Haldane, J. S. B. *The Causes of Evolution*.

Johnson, Edey and Ronald. *Blueprints*.

Leakey, Richard. E., and Roger Lewin. *People of the Lake*.

Lewin, Roger. *In the Age of Mankind*.

———. *Thread of Life*.

McCollister, Betty, ed. *Voices for Evolution*.

Mayr, Ernst. *One Long Argument*.

———. *Systematics and the Origin of Species*.

Simpson, George Gaylord. *The Meaning of Evolution*.

Strahler, Arthur N. *Science and Earth History: The Evolution–Creation Controversy*.

Willis, Delta. *The Hominid Gang*.

Television Channels

Discovery Channel.
Public Television Channel.
> (These television stations have quite a few programs about evolution. Sometimes other channels do also.)

Newspapers and Magazines

Some of our larger newspapers have articles about fossil finds, also how evolution affects our lives, as a story this morning tells how women would have less breast cancer if they did some of the things Stone Age women did (story in *Daily Oklahoman*, February 16, 1993).

National Geographic: Many articles about dinosaurs, early man, and evolution.